PLANT PROTECTION IN THE PACIFIC 3

...tissue culture...

Comments on Tissue Culture in the Pacific Region
and an account of Tissue Culture work
at the Institute for Research, Extension and Training
in Agriculture (IRETA), University of the South
Pacific (USP), Alafua Campus, March 1992-May
1993.

SEMISI PONE
BSc, MSc (Hons).

Acknowledgment.

I would like to acknowledge the contributions of Mr Tevita Holo BSc, MSc; Principal Plant Pathologist, MAFF, Kingdom of Tonga and Dr Mary Taylor PhD, Fellow in Tissue Culture, IRETA, USP, Alafua Campus, Samoa to this book through discussions of our work and ideas.

I would also like to acknowledge the financial contribution of the USP/EU Pacific Regional Agriculture Programme (PRAP) Project 7 to the work that I was involved in at IRETA/USP. I believe germplasm collection and storage will be one of the cornerstones of Pacific Agriculture in the future.

This book is dedicated to the memory of Mr Tevita Fa'oa Holo BSc, MSc; Principal Plant Pathologist and Head of Research, Research Division, Ministry of Agriculture, Fisheries and Forests, Kingdom of Tonga who was my Science and Research colleague from June, 1985-March, 1992. A brilliant man who contributed a lot to the development of tissue culture and agriculture in Tonga, and its MAFF research staff.

CONTENTS.

LIST OF TABLES.

LIST OF FIGURES.

Introduction.

This book is an account of my work at the Institute for Research, Extension and Training in Agriculture (IRETA), University of the South Pacific, Alafua Agricultural Campus, Apia, Samoa from March 1992 to May, 1993. I was a Fellow in Tissue Culture working mostly with the germplasm collection of sweet potato (*Ipomea batatas*) and vanilla (*Vanilla fragrans*). I also did some maintenance work with the other collections like yams (*Dioscorea spp*) and bananas (*Musa spp)*.

I had carried out several experiments to find the best media for slow growth of sweet potato to lessen our maintenance, sub-culturing work and also rapid multiplication of vanilla. Some USP member countries were interested in planting commercial vanilla plantations and we wanted to be ready to supply the planting material when they need them.

I thought that these experimental results and

my comments may be useful to other Agricultural Science and Plant Protection staff in the Pacific region, who are working with these 2 plants *in vitro*.

I will leave the literature research for NPPS staff, USP students and others who want to use my results as it will take up too much of my time and it is not my intention to prove anything other than present the results which are fairly clear.

I am also using statements from my work at SPC (www.spc.int) and various other organisations (www.fao.org) to support my discussions and explanations. I am sure they will understand, it as an aid to Scientific Research in Agriculture in the Pacific region which we all want to promote.

Tissue Culture will, increasingly, be used in Pacific Agriculture as I have predicted back in 1986. There are many benefits which are discussed in this book by myself and other organisations. The conclusion is clear.

Tissue culture will be used to mass produce planting material, like TLB recovery in Samoa. It is the most logical solution. Growers can be supplied with a large number of resistant taro cultivars to revive the Samoan Taro Industry within a short period of time.

It is being used to transfer clean planting material between countries and a strategy against microbial diseases in Plant Protection. For example, the Australian Centre for International Agricultural Research (ACIAR) has been carrying out banana research in Tonga with clean tissue cultured banana plantlets from Australia, for many years.

Storage of clean, popular and high value germplasm for research and future use is another effective Tissue Culture intervention like the Unit at IRETA/USP and SPC.

Rapid multiplication of some commercial crops like bananas and flowers like orchids depend on tissue culture. Its uses will increase not only in number but importance around the Pacific Islands.

Many modern Tissue Culture Units are being build in the Pacific like the recently completed unit in Fiji. SPC and USP are promoting regional co-operation between the Pacific Islands which will help modernize, not only the commercial sector, but the subsistence farming as well.

Improved and high value cultivars, disease resistant cultivars and export or commercial crops are just the beginning.

I have underlined the important sections on the statements I took off the internet, from SPC, FAO, IAEA, WIKIPEDIA and others to emphasize and support my explanations here. The idea is to show how Tissue Culture is used by various organisations, and how it is used outside the classrooms and laboratories of the Universities. It will be useful to students in the Pacific.

Lastly, I wish to thank all the staff at USP,

Alafua and friends in Samoa for making our stay in Apia, such a memorable time.

Chapter 1. What is Plant Tissue Culture?

Here's the definition from wikipedia, the free online encyclopedia...

'Plant tissue culture is a collection of techniques used to maintain or grow plant cells, tissues or organs under sterile conditions on a nutrient culture medium of known composition. Plant tissue culture is widely used to produce clones of a plant in a method known as micropropagation.

Different techniques in plant tissue culture may offer certain advantages over traditional methods of propagation, including:

- The production of exact copies of plants that produce particularly good flowers, fruits, or have other desirable traits.
- To quickly produce mature plants.
- The production of multiples of plants in the absence of seeds or necessary pollinators to produce seeds.
- The regeneration of whole plants

from plant cells that have been genetically modified.

• The production of plants in sterile containers that allows them to be moved with greatly reduced chances of transmitting diseases, pests, and pathogens.

• The production of plants from seeds that otherwise have very low chances of germinating and growing, i.e.: orchids and *Nepenthes.*

• To clean particular plants of viral and other infections and to quickly multiply these plants as 'cleaned stock' for horticulture and agriculture.

Plant tissue culture relies on the fact that many plant cells have the ability to regenerate a whole plant (totipotency). Single cells, plant cells without cell walls (protoplasts), pieces of leaves, stems or roots can often be used to generate a new plant on culture media given the required nutrients and plant hormones'.

●●●

Here are some pictures of what a Tissue Culture scheme and plantlets in the storage bottles look like...

Figure 1. A typical scheme of tissue culture operation
From agritech.tnau.ac, internet pictures .

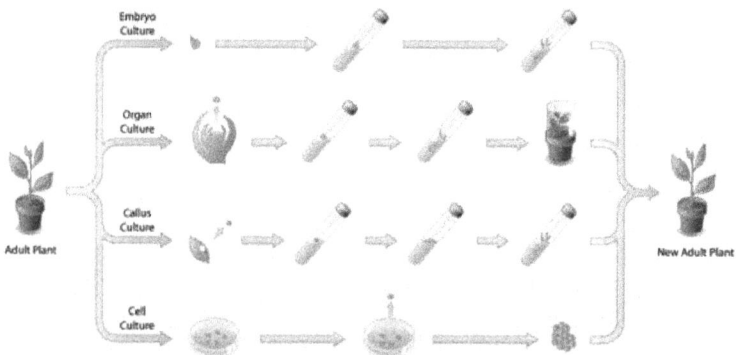

This scheme shows the steps from beginning to end starting with plant embryo, organs, callus or cells.

Meristem Culture...
Meristem culture is used in cases where a plant need to be cleaned of a virus disease. The meristem is removed and cultured because, very often, it is free of viruses. The meristem is the tip of the shoot.

TISSUE CULTURE PICTURES...

Figure 2. The plantlets below looks exactly like sweet potatoes (*Ipomea batatas*) *in vitro*. From 'internet pictures'.

Figure 3. The plantlets below are generated from "callus" which is a mass of rapidly dividing plant cells. From 'internet pictures'.

Figure 4. The picture below shows how mass
multiplication of tissue cultured plantlets looks like.
Large numbers of plantlets are mass produced inside the
storage bottles. From flora.coa.gov.tw, 'internet pictures'.

Figure 5. The picture below shows a close up picture of
what the plantlets look like inside the storage bottle.
These plants are ready for planting to the field. From the
Atlanta Botanical Garden, 'internet pictures'.

Chapter 2. The need for Tissue Culture in the Pacific Islands.

Most of the 22 Island Nations of the Pacific Region relies on agriculture for subsistence as well as economic growth. Of the 15 countries listed by the Secretariat for the Pacific Community (www.spc.int) in its economic activity chart, 5 countries listed agriculture export as "more than 50%" of the total export. Seven countries listed more than 40% and 8 countries listed more than 30%.

Amazingly, 3 countries (Vanuatu, Solomons and Kiribati) listed agriculture export as more than 80% of the total!. That is a huge portion of the income for the country!.

There are more than 9 million inhabitants in the Pacific Islands most of which live in Papua New Guinea, Solomon Islands, Vanuatu and Fiji. The percentage of these people engaged in agriculture in the Pacific is normally very high, as indicated by the agriculture export figures. In some Pacific countries as much as 90% of the population are engaged in agriculture activities, mostly subsistence farming.

These figures prove conclusively that agriculture is the most important economic activity in the Pacific Islands and it is crucial to the livelihoods and economic welfare of its people.

However, there are many problems facing both economic and subsistence agriculture in the Pacific. These include;

1. Lack of good quality planting material.

Tissue cultured clean planting material

Figure 6. Cavendish banana plantlets raised from tissue culture, ready for planting to the field. MAFF, Tonga.

can be supplied in large numbers if the facilities are available. Producing all planting material from tissue culture, avoids transfer of plant virus, bacterial, fungal diseases, insects and nematodes with the planting material.

I visited the Republic of China on Taiwan in 1986, while working for the Ministry of Agriculture Fisheries and Forests, Tonga, to get some experience in their tissue culture operation. I was surprised that is was a huge operation. All the local banana planting material for commercial plantings were completely controlled through tissue culture. The planting material, planting dates, diseases, insects and the harvest were controlled through the tissue cultured plantlets.

Figure 7. Cavendish banana planted from clean tissue cultured plantlets. Banana Research Institute, Taiwan, Republic of China.

The uniformity of the planting material, avoidance of nematodes, fungal, bacterial and virus diseases were all possible through tissue culture. All these diseases were previous problems of the Banana Industry in Taiwan, Republic of China.

2. Lack of improved or disease resistant varieties.

In the case of taro leaf blight in Samoa, the lack of resistant planting material is a major constraint to production. There are known resistant varieties of taro in the Pacific Islands. Papua New Guinea had a major breeding programme for resistant taro varieties in the early 1990s at the Lae Research Station with Mr Sim Sar and Dr Anton Ivancic implementing it.

Figure 8. Taro Leaf blight (TLB) resistant taro in the Papua New Guinea breeding programme. This picture was taken during the wet season when disease pressure is very high. Note none of the leaves show any necrotic areas or black spots associated with TLB. Lae Research Station. Papua New Guinea.

It would be sad if those resistant taro varieties are not made available where they are needed, like in Samoa for example.

I was fortunate enough to visit the Lae Research Station many times during the dry and wet season to observe the taro as well as participate in the tasting panel, trying out the improved and blight resistant taro varieties.

If the 2 regional Tissue Culture Labs at USP, Alafua and SPC, Suva are given some of these resistant varieties in the name of regional co-operation, it would be of great benefit to all countries affected by TLB.

The regional solution would simply to collect resistant cultivars or varieties of taro in the Pacific Islands and store them at the regional facilities where large numbers can be supplied to all members of USP and SPC.

Criticism has been made of the duplication of efforts in the two regional facilities but they are needed in case of hurricanes destroying one.... and the other can continue. It is highly unlikely that a hurricane can wipe out both facilities at once. Perhaps if money is available they can add a third germplasm storage facility at the Coffee

Research Station in Papua New Guinea or similar Institute in the region.

Samoa's taro leaf blight (TLB) problem can be solved easily if there is a political will to do so. Resistant varieties are already available. The next step would be to test them in large areas of Samoa to observe the effect of the TLB and yield. The locals can pick and choose the varieties they like. Multiplication of taro plantlets in millions can be done easily through tissue culture. The idea is to expose large numbers of the resistant varieties to high disease pressure in the field and let natural selection do the job. I have a feeling that testing for resistance in the breeding plots is not enough to make decisions on. It is also rather slow, as 20 years have now passed since the introduction of TLB into Samoa and resistant varieties are still not widely available to Samoan taro growers.

3. Lack of salt tolerant varieties in atolls.

Taro and giant taro grown for local

consumption, in atolls, can be tested for salt tolerance. Salt tolerance may even be induced in culture for such use. This may be useful in atolls where sea level rise is a problem like in Tuvalu and Kiribati.

4. Problems of spreading disease with planting material.

In cases where the planting material helps to spread the disease, tissue culture may be the solution. Planting material can be cleaned in culture, mass produced then distributed to growers in the islands.

This may be helpful in the case of vanilla, yam, taro, cassava and bananas.

Kava (*Piper methysticum*) remains a problem for Tissue Culture experts. No one has cultured kava successfully, *in vitro,* yet.

Forest trees and rare plants and orchids can also be cultured. Any agricultural commercial operation in the Pacific Islands can be supported by regional facilities.

...TO SUPPORT THIS ARGUEMENT I HAVE COLLECTED STATEMENTS FROM ORGANISATIONS INVOLVED IN TISSUE CULTURE....

BOTANICAL GARDENS CONSERVATION INTERNATIONAL (BGCI). The global network...

In vitro Storage Methods.

' The storage of germplasm in laboratory conditions (*in vitro*) is specially suited for the long-term conservation of recalcitrant species and vegetatively propagated species. They can be stored at low temperature under slow growth conditions or cryopreserved in liquid nitrogen at -196°C. Cryopreservation has so far been successful with only a relatively few species but is a very promising development for long-term storage. The main limitation of *in vitro* storage is the need for special equipment, techniques, trained staff and facilities. The cryopreservation of orthodox seeds represents an advantage over low-temperature storage at -20°C, both economically and in terms of viability, as liquid nitrogen is a relatively inexpensive cryogen and the seeds retain the same viability as they had immediately before storage. Therefore regeneration costs are lowered and viability testing is reduced'.

Ministry of Agriculture, Nigeria.

1000 FARMERS RECEIVE 32,000 HYBRID PLANTAIN /BANANA VARIETIES.

' The Federal Government says that building capacities of Nigeria farmers in modern agriculture is fundamental to the actualization of the nations self-reliant and sustainable food security/sufficiency.
To this end, well-over thirty two thousand (32,000) disease free improved varieties of plantain/banana nurseries/suckers have been distributed to about one thousand (1000) prime farmers including women and youth entrepreneurs in four Pilot States in the Federation'.

My Comment;

This is what needs to happen in Samoa. Large numbers of resistant taro cultivars should be distributed to taro growers on a regular basis for planting and assessment of their performance under field conditions. Otherwise, the exercise is simply academic with no benefits to the growers whatsoever.

Ministry of Agriculture, Nigeria...continued.

The Minister for Agriculture and Rural Development, Dr. Akinwumi Adesina announced this through the Deputy Director, Horticulture Transformation Value Chain and the National Project Co-coordinator, Food and Agriculture Organization (FAO-UN) in the Ministry, Mr. Michael Kanu during a-six day training of Trainees workshop for farmers held in Abia State Capital, Umuahia, recently. The workshop was entitled "strengthening plantain/banana production in Nigeria for Domestic Consumption and Export".

The Minister expressed Nigeria's profound gratitude to FAO-UN Country Representative in Nigeria, Dr. Louise Sethswealo for supporting the campaign for global food security especially, in Nigeria and urged it not to relent.

Earlier, the FAO-UN Country Representative who spoke through the Desk Officer, Technical Cooperation Project, FAO-UN, Mrs. Adeola Akinrinlola, observed that the workshop was organized to acquaint farmers in Umuahia, Abia State with the latest plantain/banana production techniques and post harvest technologies in order to make farming business a profitable venture.

The FAO-UN country representative disclosed that in a bid to accomplish the UN Agricultural support mandate to developing countries, improve in the technical skills of farmers to overcome production constraints and to establish pilot field planting nurseries/suckers as a source of Tissue Culture planting materials beyond the project life time.

FAO-UN provided a funding grant of $ 477,000 Dollars (N77m) to Nigeria`s FMARD to compliment the shortfall in her budgetary provision of $1.4m towards funding this project.

NOTE...ON TLB TISSUE CULTURE.

ACIAR had approved $AUD 1 million in 1995-1996 for the TLB tissue culture cleaning and mass multiplication of resistant cultivars from Papua New Guinea, at the Queensland University of Technology, for distribution in Samoa. Unfortunately, I left SPC in June 1996, before its implementation and recommended to ACIAR that it should be a national project rather than a regional one. I have read some of the reports from a following ACIAR TLB project in Samoa, but it appears that our plans for mass multiplication and distribution of resistant taro cultivars from PNG did not materialize.

LOW COST OPTIONS FOR TISSUE CULTURE TECHNOLOGY IN DEVELOPING COUNTRIES...
International Atomic Energy Agency, VIENNA, 2003...

Tissue culture technology is used for the production of doubled haploids, cryopreservation, propagating new plant varieties, conserving rare and endangered plants, difficult-to-propagate plants, and to produce secondary metabolites and transgenic plants. The production of high quality planting material of crop plants and fruit trees, propagated from vegetative parts, has created new opportunities in global trading, benefited growers, farmers, and nursery owners, and improved rural employment. However, there are still major opportunities to produce and distribute high quality planting material, e.g. crops like banana, date palm, cassava, pineapple, plantain, potato, sugarcane, sweet potato, yams, ornamentals, fruit and forest trees.

The main advantage of tissue culture lies in the production of high quality and uniform planting material that can be multiplied on a year round basis under disease free conditions anywhere, irrespective of the season or weather.

However, the technology is capital, labor and energy intensive. Although labor is cheap in many developing countries, the resources of trained personnel and equipment are often not readily available. In addition, energy, particularly electricity, and clean water are costly.

The energy requirements for tissue culture technology depend on day temperature, day length and relative humidity, and they have to be controlled during the process of propagation. Individual plant species also differ in their growth requirements. Hence, it is necessary to have low cost options for weaning, hardening of micropropagated plants and finally growing them in the field.

My Comment.

The Tissue Culture Unit in the Research Division, MAFF, Tonga in 1991 was build by the maintenance staff. Most of the materials were bought locally, except the lamina flow and equipment which were imported. Mr Tevita Holo, simply worked with the maintenance staff on what is required and the plan and leave them to do it. The result was an excellent working Tissue Culture Lab build and funded locally. A local girl was trained on how to carry out sub-culturing and maintenance of the plantlets. It was used for the MAFF/ACIAR Banana Improvement Project. In my opinion, given this experience, Tissue Culture facilities, equipment and expertise do not need to be imported or expensive.

Chapter 3. Facilities for Tissue Culture in the Pacific Region.

There are already suitable facilities in the region. I have worked in both the Tissue Culture Labs at the University of the South Pacific (USP) in Samoa and SPC in Suva. Both are suitable for small experiments and supplies of small numbers of plantlets for research and hurricane recovery in PIC members.

SPC Tissue Culture activities...

SPC conserves the largest taro collection in the world at its Centre for Pacific Crops and Trees (CePaCT) based in Suva, Fiji. Using tissue culture, this collection is established largely from past and ongoing taro improvement projects, including global crop regeneration projects funded by the Global Crop Diversity Trust. SPC utilises these collections for regional and global projects to benefit all countries, using agreements already established under the UN Food and Agriculture Organization's International Treaty on Plant Genetic Resources for Food and Agriculture. The first regional taro tissue culture genetic resources programme started under the EU Pacific Regional Agriculture Programme in the 1980s and then the AusAID Taro Genetic Resources Conservation and Utilisation project. The TaroGen project was initiated in 1995 in collaboration with the Samoa Ministry of Agriculture and Fisheries (MAF) and the University of the South Pacific (USP).

From www.spc.int

Tonga's tissue culture facility for multiplication of plantlets was used in the late 1980s and early 1990s for a joint banana research/improvement project with the Australian Centre for International Agriculture Research (ACIAR).

Figure 9. Tissue Culture facility at the Coffee Research Station, PNG.

It was also used for vanilla research and multiplication. The Tissue Culture Lab at the Coffee Research Institute in Papua New Guinea was the biggest in the region in the late 1990s.

I visited their facility in 1995 to assess its suitability for propagating large numbers of resistant taro plantlets for Samoa to solve the TLB problem, but did not go ahead with it due to some logistical problems of transport between the Institute, PNG and Samoa.

Fiji and other countries in the region have their own facilities for local use. It may be necessary to organize all the facilities in the region to operate together a network for exchanging improved or resistant germplasm of popular crops for the benefit of the Pacific region.

...I HAVE INCLUDED THESE STATEMENTS TO EMPHASISE THE POINT OF THIS CHAPTER...

SPC PROGRAMMES...from www.spc.int

The first regional taro tissue culture genetic resources programme started under the EU Pacific Regional Agriculture Programme in the 1980s and then came the 1993 TLB outbreak in Samoa, which triggered the SPC AusAID Taro Genetic Resources Conservation and Utilisation (TaroGen) project. The programme activities were sustained by SPC until the <u>The TaroGen project was initiated in 1995 in collaboration with the Samoa Ministry of Agriculture and Fisheries (MAF) and the University of the South Pacific (USP),initiation of the Taro</u>

improvement programme, funded by AusAID, the Australian Centre for International Agricultural Research (ACIAR) and the EU.

The past and ongoing SPC taro projects aim to restore Samoa's food security through taro improvement programmes, conserve the Pacific taro genetic resources, develop a food-secure Pacific and prepare other Pacific island countries that are TLB free, including Fiji, Tonga, Cook Islands, Niue and Vanuatu, should the blight hit their shores.

SPC conserves the largest taro collection in the world at its Centre for Pacific Crops and Trees (CePaCT) based in Suva, Fiji. Using tissue culture, this collection is established largely from past and ongoing taro improvement projects, including global crop regeneration projects funded by the Global Crop Diversity Trust (the Trust).

All new taro varieties generated out of SPC's regional donor-funded taro breeding programmes are for the benefit of SPC member countries, project partners and the global community. SPC's taro collection conserved at CePaCT has been internationally recognised, and maintenance of the collection is supported by the Trust.

SPC Ministers and Heads of agriculture and forestry have endorsed SPC taro collections to be part of the global collection under the auspices of the Treaty.

TaroGen project, and is still supporting breeding programmes based in Samoa and Papua new Guinea. Other ongoing breeding programmes are in Cook Islands, Fiji and Tonga, supported by the AusAID International Climate Change Adaptation Initiative (ICCAI).

In addition to TLB, climate change is a global threat to food security and emerging pests and diseases, such as the *Bogia phytoplasma* coconut disease. The global taro breeding programme involving over 20 member countries, including the Pacific, is implemented by SPC through the EU International Network for Edible Aroids (INEA). <u>All new taros coming out of these regional and global breeding programmes will be shared by all members through SPC.</u> The breeding programme in Samoa has generated seven to eight breeding cycles and produced new taro varieties, largely using material from the Pacific and Asian countries provided by SPC. Samoa 1 and 2 varieties were selected from progenies of cycle 5 breeding.

Cycles 6, 7 and 8 taro varieties are still undergoing research evaluation under SPC ACIAR Pacific Agribusiness Research for Development Initiative (PARDI) – Developing a clean seed system for market-ready taro cultivars in Samoa – involving the Ministry of Agriculture, the Scientific Research Organisation of Samoa and the Samoa Farmers' Association, in collaboration with the University of the South Pacific.

<u>Some of the taro varieties utilized earlier in</u>

the SPC Samoa-based breeding programme in collaboration with Samoa MAF and USP were taro varieties generously provided by other countries through different donor-funded projects. Amongst those early taro varieties introduced were the popular Talo fili (PSB-G2) from Philippines, Palau varieties (Palau 1–20), including the popular Talo polovoli (Ngerruuch), Talo tie-dye (Toantal) and Talo fai-luau (Pwetepwat) from the Federated State of Micronesia. Then in the late 1990s, new taro varieties were released by the Samoa MAF. These were locally known as Talo Suga, Seu, Asu, and were cycle 2–4 varieties under the TaroGen project, utilising existing diversity at the time.

In 2003, SPC sourced and provided more new resilient Asian taro lines from the EU Taro Network for South East Asia and Oceania for the breeding programme and, as a result, more vibrant taro varieties were produced, commonly known as Talo ta'amu and or Talo laui'ila in Samoa, as they do resemble *Alocasia* (ta'amu).

The breeding programme should never be stopped, as strains of *Phytophthora colocasiae* that caused the TLB disease found in other countries in the world might be more aggressive or severe than the strain found in Samoa and the Pacific. For this reason, SPC keeps on sourcing new tolerant taro varieties from within and outside the Pacific to be incorporated into the breeding programmes based in Samoa and other Pacific Island countries.

The main themes of the recent Treaty meeting held at SPC CePaCT in December 2013 in Suva, Fiji were: No country is self sufficient in plant genetic resources and countries inter-depend on one another. These themes promote the importance of sharing plant genetic resources for global food security as the prime aim of the Treaty. Already 131 countries are parties to the Treaty, including France and some Pacific countries (Australia, Cook Islands, Palau, Fiji, Samoa and Kiribati).

SPC acknowledges all stakeholders, development and project partners, farmers and taro breeders involved in these breeding programmes, including, in particular, Moafanua Tolo Iosefa, SPC/USP Samoan taro breeder, who is a key person behind the success of the taro project and the generation of the latest taro varieties – Cycles 5 to 8 – which offer so much good taro diversity now available in Samoa, the Pacific and the global community.

My comments...

Obviously there is a lot of work going on since the destruction of the Samoan Taro Industry by TLB in 1993 which are highly commendable. Everyone involved should be congratulated. However, as I mentioned before...it would be advantages to distribute large numbers of each resistant variety to farmers to grow. The natural selection by the TLB, the environment and farmers themselves will select the best resistant lines. I believe it would also speed up what has been or appear to be a painfully slow process.

The original plan with ACIAR in 1995, when I was the SPC Plant Protection Advisor, was to clean resistant taro varieties from Lae Research Station, PNG and multiply them to large numbers through tissue culture at the Queensland University of   Technology then distribute them to farmers in Samoa in 1996-1997. I believe it would have been a success given the Plant Breeder at Lae Research Station, Dr Anton Ivancic was getting a lot of success with his resistant breeding taro lines.

Breeding and testing the successful lines in large numbers in large areas in Samoa will expose them to TLB and the disease pressure will do all the rest.

Small plots in the Plant Breeders research gardens may not give the Plant Protection staff a clear picture of the resistance level of each breeding line without exposure to high disease pressure in large numbers of farmer's fields. That should be the next step in the breeder's scheme.

Otherwise the export of taro from Samoa
may not be a reality in the near future.

Chapter 4. Storage of popular Pacific crop germplasm for the future.

The Tissue Culture activities have been ongoing at the 2 regional laboratories (USP and SPC) for many years before I worked there. When I was involved, there were large numbers of taro varieties (100+), sweet potatoes (100+)   and lesser numbers of cassava, yam and other crops at the IRETA/USP Tissue Culture Lab and also at SPC.

In this chapter, I will include statements from Fiji, SPC and Africa on the importance of Tissue Culture germplasm collections and biosecurity to Pacific Agriculture.

The aim is to emphasize its relevance to the Pacific Islands.

NEW TISSUE CULTURE LABORATORY IN FIJI AND ASSOCIATED WORK.
...from www.spc.int

The commissioning of Fiji's tissue culture

facility last Thursday by the Minister for Agriculture, Fisheries and Forests, Rural and Maritime Development and National Disaster Management, Mr Inia Seruiratu, marked another milestone in the partnership between Fiji's Ministry of Agriculture, the Secretariat of the Pacific Community (SPC) and Australian Aid.

The new tissue culture laboratory was built with assistance provided by the Fiji government and the Australian aid programme through SPC.

The Fiji government provided a total of $115,745 from the Rural and Outer Islands Programme to build the laboratory, while Australia provided $100,000 through SPC for equipment, consumables, glassware, nurseries, laboratory staffing and training to enhance technical capacity.

The new facility was built following an assessment carried out after the 2009 flood that devastated agricultural production in many Fiji communities, destroying much of their crops. The assessment identified the need to have a ready source of planting material for food crops after major disasters.

The laboratory will provide supplies of tissue culture plants to established nurseries around the country where they will be kept in readiness for communities affected by disasters.

By using tissue culture technology, the Ministry of Agriculture can continue to produce clean disease-free planting material as required to sustain increased production levels. SPC's Centre for Pacific Crops and Trees (CePaCT) has provided the ministry with planting material for improved varieties of several food crops, including nutrient-rich, climate and disease-resilient crops.

SPC works in partnership with Pacific countries and territories, CGIAR (Consortium of International Agricultural Research Institutes) and the Global Crop Diversity Trust to source and acquire new crop diversity for the region. This sharing of genetic resources is facilitated by the International Treaty on Plant Genetic Resources for Food and Agriculture, of which Fiji is a contracting party.

SPC also works closely with the Biosecurity Authority of Fiji to fast-track the release of suitable new crop varieties to Fiji farmers:

Over 2,000 plantlets of more than 200 varieties including climate-resilient banana, cassava, sweet potato and taro have been provided.

Around 30 sweet potato accessions have been evaluated, and four varieties that have shown drought tolerance will be released soon to farmers.

Twenty lines of taro from Samoa, PNG, Philippines and Malaysia that are tolerant to taro leaf blight are under evaluation and are also included in the current taro breeding programme. The work will help ensure Fiji is ready if taro leaf blight crosses its borders.

Eleven varieties of Irish potato tolerant to bacterial wilt are being evaluated. Two show promise and will be further evaluated by the ministry in Sigatoka. If they are suitable, they will provide a substitute for some imports.

SPC provided yellow to orange fleshed sweet potato varieties high in carotenoids and purple fleshed varieties high in antioxidants, as well as taro and cassava. These crop varieties give consumers a wider choice of nutritious food.

The ministry and the Pacific Breadfruit Project are conducting field research on over 300 tissue culture plantlets of breadfruit.

Seven banana varieties resistant to black leaf streak have been evaluated, and three FHIA varieties originating from the Honduras Foundation for Agricultural Research have been recommended by the ministry based on their performance and taste.

The new laboratory will also be used to support other sectors such as floriculture, horticulture and forestry. It will thus play a vital role in supporting Fiji's food security and economy.

The opening was also attended by the Solomon Islands High Commissioner to Fiji, HE John Patteson Oti, and Valerie Saena Tuia, Genetic Resources Coordinator, representing SPC's Land Resources Division, scientists, farmers, flower growers and other private sector representatives.

•••

Two Pacific Island government staff members undertook attachments at the Secretariat of the Pacific Community (SPC) Land Resources Division in Suva, Fiji, earlier this year. The trainees, from Vanuatu and Tuvalu, spent time at SPC's Centre for Pacific Crops and Trees (CePaCT) and its Biosecurity and Trade Support (BATS) team.

Merriam Seth Toalak, Acting Director for Biosecurity Vanuatu, undertook a five-month attachment as part of the requirements of the 2013 leadership and professional development placement under the Greg Urwin Award scholarship from September 2013 to February 2014.

Ms Toalak spent three months with the BATS team on the Biosecurity Information Facility (BIF) system. The BIF system has four components: (i) biosecurity legislation, (ii) biosecurity manual, (iii) biosecurity database, and (iv) biosecurity website. The operation manual, database and websites all link to the biosecurity legislation, which is in progress in Vanuatu.

Biosecurity Vanuatu underwent restructuring in 2012; hence there is a need to revisit and complete work on the country's BIF. Consequently, this was the focus of the leadership attachment with BATS. The development of the BIF operational manual began from the online regional model with reference to the national quarantine manual.

The same applies for the BIF database and website.

At CePaCT, she gained hands-on experience with tissue culture techniques and virus indexing, including practical indexing using molecular diagnosis and other diagnostic techniques on known plant viruses. Using polymerase chain reactions, she tested taro plantlets for the presence of taro viruses such as taro badna virus. She also tested for cucumber mosaic virus on kava. Use of these methodologies complemented her biosecurity work, especially the transfer of crop varieties across international borders.

According to Ms Toalak, 'Being placed with CePaCT with the aim of understanding the concept and skills involved in molecular and serological diagnosis, is an excellent way of translating many of the theoretical concepts studied at university into a "real-world" scenario.'

The European Union-funded SPC Global Climate Change Alliance: Pacific Small

Islands States (GCCA: PSIS) project supported the second trainee, Epu Falenga, from the Department of the Environment in Tuvalu, who spent two weeks at SPC CePaCT in February 2014. The GCCA: PSIS project in Tuvalu is trialing integrated agro-forestry systems. The project involves growing 'climate- resilient' crops under the existing coconut trees, thereby maximising the land available for agriculture and building resilience to climate change. One aspect of the project in Tuvalu is research into the effectiveness of the climate-resilient crops produced by SPC CePaCT. These crop varieties can tolerate salinity, high rainfall and temperature extremes, making them able to withstand the projected impacts encourage atoll nations.

According to Mr Falenga, 'We are trying to encourage every household on the island to grow crops between coconut trees. The training has helped me understand and promote at home CePaCT's valuable contribution to enhancing food security in the Pacific region.' He said that he was impressed with the vision and dedication of CePaCT toward conserving, developing and distributing improved varieties of crops. Swamp taro, a traditional staple crop of Tuvalu is also conserved at CePaCT.

The training provided by the CePaCT team was led by Amit Sukal, Plant Virus Diagnostic Officer (virus indexing) and Logotonu M. Waqainabete, Curator (tissue culturing).

CePaCT continues to respond to requests for attachments under various country projects to support human resource capacity development and technology transfer for agricultural improvement.

Cryopreservation is storage at ultra low temperature (-196°C), usually in liquid nitrogen. All cellular division and metabolic processes stop at this temperature, so plant material can be stored without alteration or modification for an unlimited period of time in a small volume, protected from contamination and requiring very little maintenance. Cryopreservation currently offers the only safe and cost-effective option for the long-term conservation of genetic resources of vegetatively propagated species.

Cryopreservation protocols have been developed for numerous species from both temperate and tropical origins. Once techniques are optimised, their application in genebanks ensures safe and efficient long-term storage of the germplasm of species that are important for future food and nutritional security.

However, the techniques involved in cryopreservation, in particular, meristem culture, can take some time to perfect. The meristem, or shoot-tip tissue, used for cryopreservation has to be 0.8–1 mm in size; in excising tissue this small, there is always a risk of damaging it, thus killing the plant, which will affect the regeneration rate (the number of plants obtained from the cryopreserved meristems). High regeneration rates are required, both to cover the losses that might occur but, importantly, to ensure that samples can be taken to monitor viability and to provide enough plants per accession for utilisation.

A skilled technician takes approximately one hour to excise four meristems and, ideally for each accession, there should be three independent repetitions of about 55 meristems per accession. This illustrates the significant labour input required to cryopreserve collections. The more skilled the excision, the better the regeneration rate, once the cryopreservation protocol has been optimised.

Another challenge faced in this particular research project is access to liquid nitrogen (LN), which has to be purchased from a source approximately 30 miles away from Narere, where CePaCT is located, and depends on the availability of the LN supply from overseas.

CePaCT's edible aroid collection is significant, consisting of a large *in vitro* collection of taro (*Colocasia esculenta*) of 1050 accessions mainly from the Pacific region, but also some from Asia.

The Asian collection has been expanded with the recent addition of accessions from Indonesia; further additions from other Asian countries such as Thailand and Philippines are planned. Collections of three other edible aroids, namely *Alocasia mycorrhizas, Cyrtosperma merkusii* and *Xanthosoma sagittifolium* are being established, for example, the collection of *C. merkusii*, now comprises 50 accessions.

My Comments.

The Pacific Plant Protection Organisation was established in 1994 by resolution of the 34th South Pacific Conference. It has been under discussion for 8 years previously and the resolution was finally accepted in 1994. I had worked on the resolution document with the FAO Legal Advisor, Mr Richard

Stein, and the New Zealand Foreign Affairs Department with support from the Director of Programmes, Mr Poloma Komiti, Forum Secretariat staff and others before it was submitted at the 34th SPC Conference.

It has now become a very important organizational tool for the Pacific Region in matters of biosecurity and facilitates trade in tissue cultured germplasm and related work. It is the regional plant protection organisation (RPPO) for the Pacific representing the islands at important global meetings. I had represented the PPPO at RPPO meetings at FAO, Rome every two years and also the Biosecurity expert panel (CEPM).

I have included statements from the www.spc.int website here to emphasize the importance and relationship between the Pacific Plant Protection Organisation and Tissue Culture in the region.

Most of that work, I had initiated myself in 1994-1996.

Biosecurity and Trade.
(source-www.spc.int)

Biosecurity is a strategic and integrated approach to analyzing and managing relevant risks to human, animal and plant life and health and associated risks to the environment. Interest in biosecurity has risen considerably over the last decade in parallel with increasing trade in food, plant and animal products. There is more international travel, new outbreaks of trans boundary disease affecting animals, plants and people and heightened awareness of biological diversity. Greater attention to the environment and the impact of agriculture on environmental sustainability must be maintained.

Pacific Island Countries and Territories need to position themselves to take advantage of trading opportunities, while protecting their natural resource base from potential risks.

The PPPO council meets every three years but the PPPO Executive committee comprising of representatives from the three different geographical subregions; Melanesia, Micronesia and Polynesia plus Australia and New Zealand meet annually.

The PPPO executive committee members are Cook Islands, Tonga, New Caledonia, Solomon Islands, Federated States of Micronesia and Nauru.

The Secretariat of the Pacific Community hosts the PPPO Secretariat as it is the regional organisation that hosts all member countries and is involved in providing assistance in plant protection and quarantine to member countries since its inception in 1947. Non-IPPC contracting parties in the region such as the French, US and New Zealand Territories are also invited to participate in PPPO meetings for their information and to implement biosecurity standards. In the standard setting process such as the draft ISPM workshops member countries meet to discuss and endorse standards and discuss ways to implement the standards. The PPPO secretariat also facilitates information exchange amongst its member countries.

The activities implemented by the Biosecurity & Trade Support (BATS) Team are based on the SPC Land Resources Division Strategic Plan (2009-2010).

Objective number 3 on improved biosecurity and increased trade in agriculture and forestry products.

Objective 3.1: National capacity to comply with international and other relevant standards strengthened:
To facilitate trade in agricultural and forestry products, PICTs must meet a variety of international sanitary and phytosanitary (SPS) and food safety standards. Biosecurity & Trade Facilitation plays a crucial role in assisting countries to comply with international standards through training and capacity building to enhance PICTs ability to take advantage of trading opportunities.

The BATS group provides technical services to the PICTs in the following areas:

- upgrade and maintain national border biosecurity (quarantine) services and biosecurity related trade services including harmonization of laws and procedures
- compliance with international standards
- capacity building in trade facilitation procedures
- <u>ensure safe movement of germplasm in the region</u>
- strengthen border operations by supplying facilities, training in the use of new improved facilities and continued support for training in inspection and surveillance as the first line of defence to minimize incursions.

Objective 3.2: National capacity to increase domestic and export trade developed and strengthened:

Increasing international trade in agricultural and forestry products relies on obtaining market access approvals from importing countries. BATS has been working to build capacity of PICTs to complete market access requests as part of the training provided by its Import and Export Technology Centre (IMPEXTEK). These activities are vital if PICTs are to realize the benefits of international trading opportunities.

The BATS group achieves the above objective by assisting PICTs in the following activities by assisting PICTs
- to conduct import risk analysis
- develop import protocols
- develop trade access submissions and
- to comply with export protocol and certification based on international standard, regional or trade partner procedures.

Objective 3.3: Sustainable and viable post-harvest technologies developed and promoted:

Post harvest handling and value-adding methods are important to meeting the quality requirement of the market and the quarantine requirements of importing countries. The small economies of PICTs preclude competing on cost grounds with many other competitors, but the development of innovative post-harvest technologies and value adding methods can assist countries in developing niche products where they have comparative advantages. BATF continue to support research activities to develop and evaluate appropriate post-harvest technologies by;

- facilitating economic analysis
- marketing research and evaluation of national capacity to maintain continuity of commodity supplies for domestic or of potential export commodities
- collaborating with other LRD groups and stakeholders develop programmes focusing on supply chains to raise the quality and reduce post-harvest losses and to add value to agricultural and tree products for domestic and export trade
- consulting with traders, trade partners and producers , facilitate and support the development of appropriate pre-shipment treatments and handling to increase shelf

life, reduce spoilage and compliance with the sanitary and phytosanitary requirements of trade partners.

Output 3.4: Improved information available on plant and animal health status

Accurate information on plant and animal pest status is a precondition for trading in agricultural and forestry products. The Pacific Island Pest List Database (PIPLD) provides support for PICTs wanting to access overseas markets and maintain existing market access. These regional databases will further enhanced by BATF direct participation on the various global compendia produced by the Commonwealth Agricultural Bureau International (CABI) Trade has an important role to play in supporting economic growth in PICTs with trade in agricultural and forestry products. It must be supported by effective biosecurity support services to enable PICTs to submit market access requests based on valid scientific evidence.

BATS supports the capacity of PICTs

- <u>to comply with the requirements of importing countries and ensure imports comply with their own requirements and thus reducing the risks of introducing new pests and diseases due to increased trade and movement of people</u>
- maintain efficient databases for rigorous and up-to-date, readily accessible records of the complete list of all pests and diseases of plants and animals essential for countries of the Pacific to produce credible quarantine import protocols (based on IRAs) and to make acceptable market access submissions for exports
- assist PICTs develop national biosecurity facility based on the regional model to enable improved biosecurity operations and risk based decision-making on biosecurity issues and
- maintain a biosecurity alert and information service to rapidly inform PICTs of imminent pest and disease threats and to provide precautionary technical advice.

African Crop Science Journal
AFRICAN CROP SCIENCE SOCIETY

Tissue culturing has become a routine method for propagating plants in high technology laboratories. The cost of production using conventional tissue culture is, however, high for most of the countries in the sub-Saharan Africa.

In this study, we evaluated a micro-propagating protocol for local banana (*Musa* spp.) in Kenya as an alternative to reduce the unit cost of tissue culture micropropagation. Matrices were satisfactory and comparable to the gelling agents. Glass beads were, however, the best matrix in shoot multiplication. Use of support matrices, locally available macronutrients, micronutrients, sugar, equipment and facility reduced the cost of consumable material for banana tissue culturing by about 94%.

Putting into account energy, labour and capital investments, the cost dropped from approximately US $ 1.5 to 1.0* per plantlet.

Contamination was not observed when the media and equipment were sterilized using a pressure cooker instead of an autoclave. Use of plastic syringes instead of glass cylinders and micropipettes, to measure volumes reduced the cost of the equipment by 96%.

The risk of damage and loss due to breakage was eliminated compared to the use of glassware equipment. Shoots were rooted when they were transferred to Murashige and Skoog (MS) medium supplemented with 1 mg l^{-1} napthaleneacetic acid (NAA) or 1 mg l^{-1} Anatone.

Acclimatized plants were successfully transplanted and established in the field. There is potential for use of locally available low cost resources as alternatives to the conventional costly laboratory resources.

---

* - Mass multiplication of banana plantlets in other countries should be cheaper than the $US1 each quoted above.

Chapter 5. Development of a long-term storage system for sweet potatoes (*Ipomea batatas*).

In this chapter, I will present data that I have kept with me over the last 23 years...of the experimental work I did at the USP/IRETA Tissue Culture Unit. I simply did not have time to write them up before. I hope that Agricultural Science, Plant Protection and Tissue Culture staff in the Pacific Islands and other countries will find them useful.

In March 1992, I joined the Institute for Research, Extension and Training in Agriculture (IRETA), University of the South Pacific, Alafua Agriculture Campus in Apia, Samoa. My position was a Fellow in Tissue Culture in the Pacific Regional Agriculture Programme (PRAP) Project 7. It was funded by the European Union as part of its assistance to ACP countries (Africa, Caribbean, Pacific) under the Lome Convention.

My job was to maintain the sweet potato

(*Ipomea batatas*), vanilla (*Vanilla fragrans*), yam (*Dioscorea spp*) and banana (*Musa spp*) collections at IRETA/USP. I also handle all requests from member countries and supply tissue cultured plantlets of popular Pacific crops for hurricane recovery and research. My colleague Dr Mary Taylor and support staff also looked after part of the germplasm collection, such as the taro (*Colocasia esculenta*), cassava (*Manihot esculenta*) and other collection work.

All the germplasm accessions were cultured in MSO. That is Murashige and Skoog's (1964) basal medium plus sucrose, as a carbon source.

The most labor intensive part of maintaining the germplasm is the sub-culturing of the plantlets when they outgrow the container. There are hundreds of accessions in the germplasm collection. Sweet potato and taro combined were more than 300 accessions. So it is a very time consuming exercise.

Everything must be sterile, so all media,

work benches and sub-culturing tiles must all be sterilized, before the work begins.

We wanted to find a better media for storage. One that will keep the plantlets in good health while maintaining slow growth *in vitro.* This will lower the amount of sub-culturing and maintenance required. With that objective in mind I carried out a series of experiments to determine the best media and supplements for long-term storage of sweet potatoes (*Ipomea batatas*) *in vitro.*

Murashige and Skoog (MS) medium...

MS is a plant growth medium used in the laboratories for cultivation of plant cell culture. MS was invented by plant scientists Toshio Murashige and Folke K. Skoog in 1962 during Murashige's search for a new plant growth regulator. Sucrose is added to the MS and is referred to as MSO. Along with its modifications, it is the most commonly used medium in plant tissue culture experiments. Wikipedia.

Several experiments were conducted to determine conditions under which the *in vitro* growth of sweet potato (*Ipomea batatas*) can be slowed down thus saving valuable working time by reducing the amount of sub-culturing required. These conditions include the use of diuretics such as sugar alcohols (mannitol, sorbitol), nutrient and sucrose (carbon source) reduction, growth regulators such as abscisic acid and alternative carbon sources such as coconut water.

All the experiments used nodal explants from the IRETA Tissue Culture Unit of the University of the South Pacific, Alafua Agriculture Campus, germplasm collection.

All experiments used completely randomized designs with computerized statistical analysis. Treatment 1 was always used as the control throughout these experiments. Unless stated otherwise, all cultures were maintained under normal tissue culture lab conditions at 42- 48 umol $m^{-2}s^{-1}$ photosynthetic active radiation (PAR), $24\pm2°C$, and 12 hours photoperiod. Nodal explants were used per treatment and the growth indicated by plant height was recorded after 12 weeks.

The sweet potato germplasm collection.

Sweet potato is a very important staple food in the Pacific Islands, both as a subsistence and commercial crop.

One of the purposes of the collection was to store disease susceptible cultivars for future use. For example, the cultivar "Tongamai", from Tonga, has excellent eating qualities but is highly susceptible to sweet potato leaf scab caused by *Elsinoe batatas*. Perhaps one day resistant varieties of this excellent cultivar can be bred.

There was more than 100 accessions of sweet potato at the EU/IRETA Tissue Culture Unit.

EXPERIMENT 1. Effect of sucrose on the growth of *Ipomea batatas* cv IB17 *in vitro*.

In this experiment, the plant height and the number of nodes produced were used as indicators of growth.

Sucrose as a carbon source and additive to the MS promotes faster growth. The aim of the experiment is to assess the effect of various concentrations of sucrose and its effect on the growth of IB17.

RESULTS AND DISCUSSIONS.

All treatments containing sucrose grew significantly faster than the control (Treatment 1), which had no sucrose in it. In fact, there was no growth in all the explants of this Treatment 1.

The results suggest that controlling the amount of sucrose and carbon available to the plants, can control the growth of this cultivar.

The plantlets grew faster at higher concentrations of sucrose. Treatment 2, 3, 4 and 5 were all significantly better than the control in terms of growth. It suggests that the availability of a carbon source such as sucrose is very important for *Ipomea batatas* growth *in vitro*.

The higher concentrations of sucrose also performed better than the lower concentrations in terms of plant height and number of nodes. It appears that the larger amounts of carbon available enhances the growth of the plantlets.

Table 1. Effect of sucrose on the growth of *Ipomea batatas* cv IB17 *in vitro*, after 12 weeks growth (means of 16 plantlets).

Treatments	Nodes	Plant Height (mm)
1.MS	0	0 a*
2.MS+1%	10	23 b
3.MS+2%	10	45 c
4.MS+3%	13	45 c
5.MS+4%	11	58 d

*-Treatments with the same letters are not significantly different, LSD (5%) = 2.11.

The LSD is the least significant difference or the number above which the treatment is significant. For example, in Table 1 the LSD is 2.11 mm. Treatments that differ more than 2.11 mm in height are significantly different at the 5% level of significance.

All analysis of these experiments were done by the computer.

EXPERIMENT 2. Effect of reduced nutrients (MS and MSO) on the growth of *Ipomea batatas* cv TIS 5081 from Samoa (means of 8 plants).

This experiment was designed to assess the response of TIS 5081 to reduced amounts of nutrients in the media with or without sucrose (MS and MSO). The experiments were carried out to determine whether reducing nutrients, and carbon source, to the sweet potato plants is a viable method for achieving slow growth of *Ipomea batatas* cv TIS 5081 *in vitro*.

The experiments involved serial dilutions of Murashige and Skoog's (1964) basal salt medium (MS) without sucrose and with 3% sucrose (MSO), as the carbon source.

Both experiments were maintained under 42- 48 umol m^{-2} s^{-1} photosynthetic active radiation (PAR), 24$\pm$2°C, and 12 hours photoperiod. Eight nodal explants were used per treatment and the growth indicated by plant height was recorded after 12 weeks.

RESULTS AND DISCUSSIONS.

Table 2. Effect of reduced nutrients (MS) on the growth of *Ipomea batatas* cv TIS 5081 from Samoa (means of 8 plantlets).

Treatments	Nodes	Plant height(mm)
1. MS	3	5
2. 1/2 MS	3	5
3. 1/4MS	2	3
4. 1/8MS	1	2
5. 1/16MS	1	2
6. 1/32MS	1	1

Plant height and the number of nodes produced, declined with decreasing nutrient content. Growth of plants in dilutions greater than 1/4, in both experiments, were abnormal with very narrow elongated leaves and shortened internodes. The absence of a carbon source (sucrose) is evident in the growth of explants cultured in MS compared to those cultured in MSO.

Table 3. Effect of reduced nutrients (MSO) on the growth of *Ipomea batatas* cv TIS 5081 from Samoa after 12 weeks (means of 8 plantlets).

Treatments	Nodes	Plant height (mm)
1. MSO	9	23
2. 1/2MSO	7	16
3. 1/4MSO	6	11
4. 1/8MSO	7	14
5. 1/16MSO	7	10
6. 1/32MSO	7	10

However, growth of plants in the diluted media were not significantly different from those on full MS and MSO, although there is a large difference in height there was little difference in the number of nodes produced, suggesting that diluted media may result in reduced inter-nodes.

The results of these experiments suggest that reducing the nutrient content of the media may not be a suitable method for achieving slow growth of sweet potatoes *in vitro*.

The effect of reduced nutrients (both MS and MSO) were reflected in the reduced growth of the explants.

EXPERIMENT 3. The effect of coconut water (undeproteinised) on the growth of sweet potatoes using cultivar "*Amelika*", from Tonga.

Six treatments were applied with Murashige and Skoog (1964) (MS) basal salt medium with increasing concentrations of 5% increments of coconut water (CW).

Cultures were maintained under 42-48 umol $m^{-2} s^{-1}$ photosynthetic active radiation (PAR),

24±2°C, and 12 hours photoperiod.

Sixteen nodal explants were used per treatment and the growth indicated by plant height was recorded after 12 weeks.

RESULTS.

Table 4. Effect of coconut water on the growth of *Ipomea batatas*, cv *Amelika* (IB52) *in vitro* after 12 weeks of growth (means of 16 plantlets).

Treatment	Plant Height (mm)	
1. MS (Control)	0.32	a*
2. MS+5% CW	3.88	ab
3. MS+10% CW	8.38	bc
4. MS+15% CW	8.38	bc
5. MS+20% CW	12.00	cd
6. MS+25% CW	15.69	d

* - Treatments with the same letters are not significantly different, LSD (5%) = 4.79.

DISCUSSIONS.

It is clear from the significant differences between Treatments 3,4,5 and 6 that coconut water is very effective as a carbon source. Although there is no significant difference between MS and MS+5% CW, the growth is highly significant at MS+10% CW or higher concentrations of coconut water.

Coconut water is a world recognised source of many tissue culture nutrients. The combined effect of all these nutrients with the sucrose in the CW gives the plants the extra boost.

It maybe possible to use small amounts of coconut water with MS as a long term storage media for sweet potatoes *in vitro,* such as 5% CW.

EXPERIMENT 4. The effect of the sugar alcohol mannitol on the growth of Ipomea batatas cv IB32 from Papua New Guinea.

The effect of the sugar alcohol mannitol on the growth of *Ipomea batatas* cv IB32 *in vitro* was investigated. MSO was supplemented with increasing amounts of mannitol and the plant growth was observed then recorded after 8 weeks. Plantlets were kept under normal laboratory conditions at 42- 48 umol m^{-2} s^{-1} photosynthetic active radiation (PAR), 24$\pm$2°C, and 12 hours photoperiod. Eight nodal explants were used per treatment and the growth indicated by plant height were recorded after 8 weeks.

RESULTS.

Table 5. Effect of mannitol on the growth of *Ipomea batatas* cv IB32 *in vitro* after 8 weeks of growth (mean of 8 plantlets).

Treatments	Nodes	Plant height (mm)
1. MSO	6	17 ab*
2. MSO+0.2%	7	25b
3. MSO+0.4%	6	16.5a
4. MSO+0.6%	6	17ab
5. MSO+0.8%	6	17.5ab
6. MSO+1.0%	6	16.75ab

*-Treatments with the same letters are not significantly different, LSD(5%) = 4.09

DISCUSSIONS.

The results show significant differences in plant height but not in the number of nodes produced suggesting that reduction in internode length maybe the cause.

There was no observed reduction in growth corresponding to increase in the amount of mannitol in the media as expected.

Mannitol amended MS may not be a suitable storage media for sweet potatoes *in vitro.*

EXPERIMENT 5. The effect of the sugar alcohol sorbitol on the growth of *Ipomea batatas* cv IB32 *in vitro* after 12 weeks of growth.

The effect of the diuretic sugar alcohol sorbitol was investigated for its suitability as a slow growth additive to the *Ipomea batatas* storage media.

Sixteen nodal explants of cultivar IB32 from the IRETA/USP Tissue Culture Lab were used. Plantlets were observed under normal laboratory conditions at 42- 48 umol m^{-2} s^{-1} photosynthetic active radiation (PAR), 24$\pm$2°C, and 12 hours photoperiod.

The plant height as an indicator of growth was recorded after 12 weeks.

RESULTS.

Table 6. Effect of sorbitol on the growth of *Ipomea batatas* cv IB32 *in vitro* after 12 weeks of growth (mean of 16 plantlets).

Treatment	Nodes	Plant height (mm)
1.MSO	12	48.5c*
2.MSO+1%	13	47.75c
3.MSO+2%	12	48.5c
4.MSO+3%	11	36.00b
5.MSO+4%	10	22.25a
6.MSO+5%	8	14.25a

*-Treatments with the same letters are not significantly different, LSD (5%) = 7.53

DISCUSSIONS.

This experiment tested the effect of the sugar alcohol sorbitol on the growth of sweet potato. It is expected that increasing the concentration of sorbitol in the media will result in the sweet potato growth slowing down.

The growth of the explants, in terms of plant height, was significantly less in the higher sorbitol concentrations (Treatments 3, 4 & 5) than in Treatments 1 & 2, in which plant growth were not affected at all.

At 4 and 5%, the loss of apical dominance was apparent. Reduction in leaf size and internode length was also evident. After 26 weeks, the plants from treatments 4 and 5 were subcultured onto sorbitol free media where 80% of the resulting plants grew normally for up to 7 weeks observation.

This result shows that sorbitol can be used as a slow growth agent for sweet potatoes.

EXPERIMENT 6. The effect of abscisic acid (ABA) on the *in vitro* growth of *Ipomea batatas* cv Amelika from Tonga.

Two experiments were carried out. Both experiments were maintained under 42-48 umol m^{-2} s^{-2} PAR. The first experiment had a 12 hour photoperiod cycle with 18-20°C (cold room) and the second had a 16 hour photoperiod cycle with 25-26°C (warm room).

RESULTS AND DISCUSSIONS.

ABA was very effective in reducing the growth of the plants *in vitro*. Treatments with higher ABA concentrations of 1mg/l grew significantly slower than the control (MSO) and 0.1mg/l ABA (Tables 7 & 8). Thus slowing the growth can be achieved in MSO with 1-10mg/l concentrations of ABA. The differences in growth between the 2 experiments suggest that warmer temperatures and longer photoperiod promote growth of this *Ipomea batatas* cultivar.

Table 7. Effect of ABA on the growth of *Ipomea batatas* cv Amelika *in vitro* under cold conditions and short photoperiod cycle after 12 weeks (means of 16 plantlets).

Treatments	Nodes	Plant height (mm)
1. MSO	13	38a*
2. MSO+0.1mg/l	11	37a
3. MSO+1.0mg/l	4	10b
4. MSO+5.0mg/l	0	0c
5. MSO+10mg/l	0	0c

*Treatments with the same letters are not significantly different, LSD (5%) = 7.83

The differences in the number of nodes between T1, T4 and T5 was significant. There were 13 nodes on the plantlets in T1 while none or no growth occurred in T4 and T5. The differences in plant height are also very obvious. The plant height of the plants in T1 was 38 millimetres while no growth was observed in the plants in T4 and T5. There is an obvious and effective reduction in growth due to the presence of ABA.

Table 8. Effect of ABA on the growth of *Ipomea batatas* cv Amelika *in vitro* under warm conditions and longer photoperiod cycle after 12 weeks growth (means of 16 plantlets).

Treatments	Nodes	Plant height (mm)
1. MSO	18	69b*
2. MSO+0.1mg/l	17	66b
3. MSO+1.0mg/l	1	3a
4. MSO+5.0mg/l	1	3a
5. MSO+10mg/l	0	0a

*Treatments with the same letters are not significantly different, LSD (5%) = 4.14

Similar to the plants in Table 7, there is a huge difference in the number of nodes and plant height between the control, T1, T4 and T5. There is definitely a significant reduction in growth due to the presence of ABA. The differences in number of nodes and plant height between Tables 7 and 8 suggest that warm conditions and longer photoperiods promote growth *in vitro*.

EXPERIMENT 7. Effect of long-term storage media on the growth of *Ipomea batatas* cv Amelika *in vitro* under the cold and warm conditions and also short and long photoperiods.

Two experiments were carried out to determine the best conditions for long-term storage of sweet potatoes under 2 different conditions of photoperiod cycles and temperatures.

The first experiment was maintained at 18-20°C with a 12 hour photoperiod (cold room). The second was maintained at 24-26°C with 16 hour photoperiod (warm room).

Both experiments were illuminated at the low light intensity of 42-48 umol m^{-2} s^{-1} PAR.

RESULTS.

Table 9. Effect of long-term storage media on the growth of *Ipomea batatas* cv Amelika *in vitro* after 6, 12 and 20 weeks growth in the cold room (means of 8 plants).

Treatment	Nodes		
	6	12	20 wks
T1	4	8	15
T2	2	5	12
T3	2	5	11
T4	3	6	7
T5	4	12	17
T6	1	4	8

T1 = MSO
T2 = MSO+2% sorbitol
T3 = MSO+1% sorbitol
T4 = MSO+3% sorbitol + 0.5% mannitol
T5 = MSO+20ppm putrecine.HCL
T6 = MS+2%sorbitol+20ppm putrecine.HCL

Table 10. Effect of long-term storage media on the growth of *Ipomea batatas* cv Amelika *in vitro* after 6, 12 and 20 weeks in the cold room (means of 8 plants).

Treatments	Plant height (mm)		
	6	12	20 wks
T1	18	27	73c*
T2	8	10	43b
T3	6	7	40ab
T4	5	8	17a
T5	13	17	76c
T6	4	6	26ab

*Treatments with the same letters are not significantly different, LSD (5%) = 25.94

T1 = MSO
T2 = MSO + 2% sorbitol
T3 = MSO + 1% sorbitol
T4 = MSO + 3% sorbitol + 0.5 % mannitol
T5 = MSO + 20 ppm putrecine.HCL
T6 = MS + 2% sorbitol + 20 ppm putrecine.HCL

Table 11. Effect of long-term storage media on the growth of *Ipomea batatas* cv Amelika *in vitro* after 6, 12 and 20 weeks in the warm room (means of 8 plantlets).

Treatments	Nodes		
	6	12	20 wks
T1	11	*	*
T2	10	*	*
T3	3	11	19
T4	4	11	15
T5	13	*	*
T6	4	9	15

*Plants in these treatments have reached top of container before measurements were made.

T1 = MSO
T2 = MSO + 2% sorbitol
T3 = MSO + 1% sorbitol
T4 = MSO + 3% sorbitol + 0.5% mannitol
T5 = MSO + 20 ppm putrecine.HCL
T6 = MS + 2% sorbitol + 20 pm putrecine HCL

Table 12. Effect of long-term storage media on the growth of *Ipomea batatas* cv Amelika *in vitro* after 6, 12 and 20 weeks in the warm room (means of 8 plantlets).

Treatments	Plant height (mm)		
	6	12	20 wks
T1	54.75c#	*	*
T2	38.25b	*	*
T3	10 a	45	90
T4	10.50a	20	40
T5	80.50d	*	*
T6	16.25a	51	*

-Treatments with the same letters are not significantly different, LSD (5%) = 14.65.
* - Plants in these containers (100mm) have reached the top of the container before measurements were made.

Growth of the plants in the 'warm room' were generally faster than those in the 'cold room', probably because of the combined effect of higher temperatures and longer photoperiod, all other conditions being the same. Within each experiment significant differences between treatments were found.

In the warm room, the best long term storage media was treatment 4 (MSO + 3% sorbitol + 0.5% mannitol). After 20 weeks, the plants were still 50% of the container height. Loss of apical dominance became apparent at 20 weeks, otherwise the plantlets were growing normally.

The result of the experiment in the cold room was a repetition of the results in the warm room but at a slower rate. Treatment 4 was, again, the best treatment in slowing the growth of *Ipomea batatas* cultivar Amelika. Average plant height was about 20% of the container height, but apical dominance was lost much earlier at 10 weeks.

Treatments 3 and 6 were better in terms of 'normal' growth. Plants in these treatments grew normally except at a slower rate than the treatments 1, 2 and 5. After 12 weeks plantlets were only 50% of the container height. It appears that the effect of the low temperature and short photoperiod helps to slow down growth as well.

DISCUSSIONS.

The best long-term storage media from the above experiments appear to be the combination of sorbitol and mannitol (MSO+3% sorbitol+0.5% mannitol) in Treatment 4, for higher temperature and longer photoperiod.

Under colder conditions with shorter photoperiod, MSO+1% mannitol or MS+2% sorbitol+20ppm putrecine.HCL may be better. MSO+abscisic acid at 1mg/l is another excellent long-term storage medium with only minimum growth after 12 weeks. At 5 and 10 mg/l ABA induces 'dormancy' of the axillary bud which can be broken by the removal of the node into ABA free medium, where growth resumes normally.

CHAPTER 6. Rapid Multiplication of *Vanilla fragrans in vitro.*

Vanilla fragrans is an important commercial crop in the Pacific Islands. In Tonga, it is the best agricultural commodity for export. It has high international price and it can be cured and stored for many months without losing its quality. It is widely grown in all the island groups of Tonga and is the main source of income for many Tongan families.

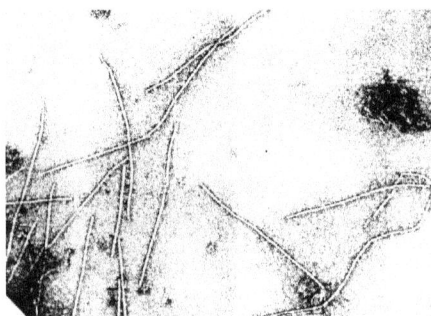

Vanilla fragrans in the Pacific Islands of Tonga and Fiji were affected by the Vanilla

Figure 10. Flexuous, filamentous particles of the VNPV from partially purified preparations of symptomatic, frozen vanilla leaves from Tonga. Electron microscope photo by S.P.Pone (approx. 10,000x).

Necrosis Potyvirus (VNPV). Tissue culture was one of the strategies used by the Ministry of Agriculture in Tonga to fight the disease by producing and distributing VNPV free vanilla plantlets to new vanilla growers.

The vanilla plantlets in the IRETA/USP Tissue Culture unit are virus free and are available to all USP member countries, upon request for research purposes or commercial ventures.

These rapid multiplication experiments aim to find the best media and supplements for rapid multiplication of *Vanilla fragrans in vitro.* They will be used by the IRETA/USP TC Unit to produce large numbers of vanilla plantlets in response to member country requests.

EXPERIMENT 1. The effect of coconut water on the rapid multiplication of *Vanilla fragrans in vitro.*

The effect of coconut water on the growth of *Vanilla fragrans in vitro* was investigated to determine the best combinations for rapid multiplication of vanilla through axillary branching and bud proliferation. The total effect of the coconut water in known to be similar to the hormone cytokinin, but because of the numerous other components of coconut water a much more enhanced effect is possible.

MATERIALS AND METHODS.

Nodal explants of *Vanilla fragrans* from the IRETA/USP TC Unit germplasm collection were used per treatment in the experiments. Sixteen explants were used per treatment. The growth as indicated by plant height was measured and recorded in millimeters. The number of nodes and shoots as an indicator of "multiplication rate" were recorded. They were sub-cultured onto MS and MS supplemented with coconut water at 5% increments up to 50% coconut water (CW) concentration.

Cultures were maintained under 42-48 umol $m^{-2} s^{-1}$ photosynthetic active radiation (PAR), $24\pm2^{o}C$, and 12 hours photoperiod.

Note: Plants in all experiments were examined throughout the length of the trial to ensure there are no problems like contamination and abnormal growth which may confound the results of the experiment.

RESULTS.

Table 13 . Effect of coconut water on the growth of *Vanilla fragrans in vitro* after 12 weeks (means of 16 plants).

Treatment	Plant Height(mm)	Nodes+shoots
1. MS	0.312 a*	0.1 a
2. MS+5%	11.00 b	1.00 ab
3. MS+10%	21.75 c	2.00 bc
4. MS+15%	23.88 cd	3.00 cd
5. MS+20%	29.50 def	4.75 e
6. MS+25%	26.38 cde	5.75 ef
7. MS+30%	25.50 cde	4.5 de
8. MS+35%	34.13 fg	5.5 ef
9. MS+40%	32.25 efg	6.5 f
10. MS+45%	32.75 efg	6.00 ef
11. MS+50%	38.75 g	9.75 g

*-Treatments with the same letters are not significantly different, LSD (5%)(nodes+shoots)=1.65, LSD (5%) (plant height)= 7.12.

Coconut water composition

Coconut water is the clear water inside the coconut. It has a high potassium content and contains antioxidants. It also contains cytokinins which promote plant cell division and growth, useful in Plant Tissue Culture. Other biologically active ingredients in coconut water include L-arginine, ascorbic acid and magnesium. Wikipedia

DISCUSSIONS.

The effect of the coconut water on the growth of *Vanilla fragrans in vitro* resulted in rapid increase in height and axillary buds with each 5% increment of CW (Table 13). The greater number of nodes obtained from axillary branching and bud proliferation means a higher number of potential plantlets at higher CW concentration. The multiplication rate of 9 plantlets per node per 10 weeks achieved in this experiment makes it theoretically possible to obtain 6561 plantlets from one node per year. Thus coconut water is a useful media for the rapid multiplication of *Vanilla fragrans in vitro.*

It is possible to respond to rapid vanilla planting programmes for disease management or vanilla commercial ventures in USP member countries.

Root development was suppressed by the MS+CW media combination. It was necessary to subculture the plantlets onto MSO to induce root growth before planting into potting mix in the screenhouse. This procedure was tested with 100% success. It was possible to get the vanilla plantlets ready for the field planting after 2-3 months in the screenhouse.

EXPERIMENT 2. The effect of sucrose on the rapid multiplication of *Vanilla fragrans* in coconut media *in vitro*.

This experiment was carried out to determine the effect of sucrose addition on the coconut media used in experiment 1. The media used was Murashige and Skoog with sucrose additive to MS (MSO) plus CW at 5% increments to 50% CW concentration. The hypothesis being tested was that the addition of sucrose will improve the multiplication rate through faster growth due to the extra carbon source (sucrose).

MATERIALS AND METHODS.

Nodal explants of *Vanilla fragrans* from the IRETA/USP TC Unit germplasm collection were used per treatment. They were subcultured onto MSO and MSO supplemented with CW at 5% increments up to 50% CW.

Cultures were maintained under 42-48 u mol $m^{-2} s^{-1}$ photosynthetic active radiation (PAR), $24\pm2^{\circ}C$, and 12 hours photoperiod.

Eight nodal explants were used per treatment. The growth as indicated by plant height was measured and recorded in millimeters. The number of nodes, plant height and root length were recorded as indicators of growth.

RESULTS.

Table 14. Effect of sucrose on the multiplication rate of *Vanilla fragrans* in MSO supplemented with coconut water after 8 weeks (means of 8 plants).			
Treatments	Nodes	Plant height	Root length
1.MSO	4.5b*	41.50abc	63
2.MSO+5%	3.25a	34 ab	27
3.MSO+10%	4.25b	51.25c	27
4.MSO+15%	4ab	33.25ab	14
5.MSO+20%	4.5b	27a	16
6.MSO+25%	4.5b	27a	9
7.MSO+30%	4.5	44bc	10
8.MSO+40%	4ab	31.75ab	6
9.MSO+50%	4ab	27a	5
*- Treatments with the same letters are not significantly different, LSD (5%) (nodes) = 0.5, LSD (5%) (plant ht) = 13.62			

DISCUSSION.

The addition of sucrose did not enhance but

reduced the multiplication rate through suppression of axillary branching. Root growth was restored by the addition of sucrose but there is an obvious decrease in length with increase in CW concentration.

Obviously, the added carbon source, sucrose, has a negative effect on the axillary buds of the vanilla plantlets, which is unexpected, probably because of increased apical dominance or root growth. It is interesting to note that there was no significant difference between T1 and T9 in terms of the number of nodes and plant height. However, there is an obvious reduction in root growth with increasing concentrations of coconut water.

EXPERIMENT 3. The effect of BAP on the growth of *Vanilla fragrans in vitro*.

This experiment was carried out to determine the potential of the cytokinin, 6-benzylaminopurine (BAP) to induce bud proliferation and rapid increase in the number of vanilla nodes and the number of plants generated for planting in the field.

RESULTS AND DISCUSSIONS.

BAP reduced the height of the plants but increased bud proliferation (Table 15). Root development was suppressed in Treatments 4, 5 and 6. At 1 mg/l, however, root development appear to be enhanced with very thick roots, twice as thick as he normal ones in the control (MSO only). Treatment 5 and 6 had increasing numbers of buds produced. However, the number of potential plantlets is still lower than the effect of CW only.

Coconut water contains nutrients that have a cytokinin effect. It seems that it is a much more effective cytokinin additive than BAP in terms of rapid multiplication of vanilla. See Table 13 and the following discussions.

Table 15. Effect of BAP on the growth of *Vanilla fragrans in vitro* after 12 weeks (mean of 8 plants).

Treatments	Plant Height	Nodes+Shoots
1. MSO	45.00 c*	4.00 ab*
2. MSO+0.1mg/l	34.75 b	3.5 a
3. MSO+1mg/l	19.50 a	2.75 a
4. MSO+2mg/l	19.25 a	5.00 bc
5. MSO+5mg/l	19.00 a	5.00 bc
6. MSO+10mg/l	16.75	5.50 c

Plant Height was measured in millimeters.

* - Treatments with the same letters are not significantly different, LSD (5%)(plant ht) = 5.35, LSD (5%)(nodes+shoots) = 1.25

Media Preparation...

All media preparation for these experiments were done by the Tissue Culture Unit Technicians. The media components are weighed, mixed with water then poured into the containers before they are sterilized with the sub-culturing tiles, wrapped in aluminium foil, in an autoclave. The bottles are cooled for the media to set, before use.

EXPERIMENT 4. The effect of napthaleneacetic acid (NAA) on the growth of *Vanilla fragrans in vitro*.

This experiment tested the effect of NAA on the growth of *Vanilla fragrans in vitro*. Plants were kept under normal lab conditions at 42- 48 umol $m^{-2}s^{-1}$ photosynthetic active radiation (PAR), 24 ± 2°C, and 12 hours photoperiod.

RESULTS.

Table 16. Effect of NAA on the growth of *Vanilla fragrans in vitro* after 12 weeks (means of 8 plants).

Treatments	Plant Height	Nodes+Shoots
1. MSO	43.88 d*	3.63 cd
2. MSO+0.1mg/l	25.88 c	2.63 b
3. MSO+1mg/l	9.25 a	1.00 a
4. MSO+2mg/l	20.75 bc	4.38 cd
5. MSO+5mg/l	23.88 c	4.50 d
6. MSO+10mg/l	16.38 b	2.88 bc

Plant height was measured in millimetres.
*-Treatments with the same letters are not significantly different, LSD (5%) (plant ht) = 7.01, LSD (nodes+shoots) = 0.88

DISCUSSIONS.

As in the experiments on reduced nutrients, the effect of NAA was evident in the reduction in plant height with increasing concentration (Table 16). Root development at 0.1 and 1mg/l were enhanced with thick

Roots twice the thickness of normal roots.

At 2 and 10 mg/l root development were retarded with very short thick roots. Roots at 5mg/l were normal. Bud development from the nodes were less in number than in experiments with NAA and NAA+BAP. It appears that the bud development is retarded at the concentrations above 5mg/l.

Subculturing...

All subculturing are done in the lamina flow cabinet in the Tissue Culture Lab. Scalpels for cutting plantlets and long forceps to transfer them into the media bottles are sterilized with dipping in 99% alcohol then flaming. The lamina flow and hands are wiped clean with 10% alcohol before and after use. Sterilized tiles are used as the "chopping board" for cutting the plantlets.

EXPERIMENT 5. The effect of the growth regulators NAA and BAP on the growth of *Vanilla fragrans in vitro*.

This experiment tested the effect of the growth regulators NAA and BAP on the growth of Vanilla fragrans in vitro.

Nodal explants from the IRETA/USP TC Unit were used with plantlets kept at normal lab conditions of 42- 48 umol $m^{-2}s^{-1}$ photosynthetic active radiation (PAR), $24\pm2°C$, and 12 hours photoperiod.

RESULTS.

Treatment Key.

1. MSO
2. MSO+0.1mg/lBAP+10mg/1NAA
3. MSO+1mg/BAP+5mg/l NAA
4. MSO+2mg/1 BAP+2mg/l NAA
5. MSO+5mg/l BAP+1mg/l NAA
6. MSO+10 mg/l BAP+0.1mg/l NAA

Table 17. Effect of BAP+NAA on the growth of *Vanilla fragrans in vitro* after 12 weeks (means of 8 plants).

Treatment	Plant ht (mm)	Nodes+shoots
T1	38.88b*	3.88
T2	21a	3.25
T3	19.5a	2.5
T4	21.88a	2.25
T5	19.13a	3.63
T6	17.38a	3.75

* - Treatments with the same letters are not significantly different, LSD (5%) (plant ht) = 7.10

DISCUSSIONS.

The strong effect of the BAP can be seen in the general reduction in plant height with increasing concentration (Table 17).

Although buds were formed there were no significant differences in the numbers of potential plantlets between all treatments. Therefore, these combinations of BAP and NAA do not appear to be conducive to the rapid multiplication of *Vanilla fragrans in vitro.*

CHAPTER 7. Testing the salt tolerance of *Xanthosoma sagittifolium*.

This experiment was aimed at determining the level of salinity that a local variety of *Xanthosoma* can tolerate. The information gathered from this experiment will then be used to design experiments on the induction of salt tolerance in this local Samoan cultivar.

RESULTS.

Table 18. Effect of artificial seawater on the growth of a local *Xanthosoma* species (means of 5 plants).

Treatments	Plant ht (mm)	Fresh wt (gms)
1. MSO	61	11
2. MSO+5%	46	5.9
3. MSO+10%	15	2.0
4. MSO+15%	6	0.35
5. MSO+20%	no growth	n/a
6. MSO+25%	no growth	n/a

DISCUSSIONS.

Growth of *Xanthosoma* was reduced with increasing concentration of artificial seawater (Table 18). At 20% and 25% artificial seawater no growth occurred. Explants in Treatments 5 and 6 turned necrotic and died within 2 weeks of subculturing. It appears that this species of *Xanthosoma* can tolerate salinity of up to 15%   as indicated by the plant height and fresh weight, although growth is gradually reduced by more than 90%.

The conclusion from this experiment is that this *Xanthosoma* cultivar can still produce leaves and corms at 5% seawater but it is reduced by 25% in height and 46% in fresh weight. At 10% seawater, production is reduced by 75% in height and 92% in fresh weight.

In terms of advice for low lying atolls in the Pacific region, it is obvious that even at 5% salt in the groundwater there is already a large 25% reduction in height and 46%

reduction in fresh weight. It may be possible to obtain edible leaves and corms but at a greatly reduced amount.

Perhaps the ideal situation would be to induce tolerance such that leaf and corm production are not affected, even at 15% salt in the ground water.

It would be interesting to test the swamp taro normally grown in atolls to see whether it is salt tolerant or not.

Xanthosoma is often used in rapid recovery from famines in the Pacific Island of Tonga because leaves can be consumed in one month and corms can be harvested in as little as 3 months while allowing the plant to grow until maturity at 8-10 months. But this production level is only possible at perfect growing conditions, after droughts.

About the author...

Semisi Pone graduated from the University of Auckland in 1985 with a Bachelor of Science. He began working as an Agriculture Officer/Plant Pathologist for the Ministry of Agriculture in Tonga in June 1985. In 1986, he began research into a new disease observed on *Vanilla fragrans* in Tonga and completed his research for a Master of Science degree at the University of Auckland in 1989. He was promoted to the position of Senior Plant Virologist in 1991 for his work on vanilla viruses(*Vanilla fragrans*), kava (*Piper methysticum*) cmv virus and squash (*Cucurbita maxima*) zymv virus. In March 1992, he joined IRETA at the University of the South Pacific, Agriculture Campus at Alafua, Apia, Samoa as a Fellow in Tissue Culture. The project he worked on was funded by the European Union in its Pacific Regional Agriculture Programme Project 7.

This book is an account of that work as well as biosecurity work he was involved in at the South Pacific Commission and the Food and Agriculture Organisation of the United Nations that relate to this work.

In May, 1993 he was appointed as the Plant Protection Advisor and Head/Co-ordinator for the South Pacific Commission Plant Protection Service based at Nabua in Suva, Fiji. He was responsible for co-ordinating the establishment of the Pacific Plant Protection Organisation (PPPO) which the 34th South Pacific Conference approved by resolution in 1994 in Noumea, New Caledonia.

He was a member of the Committee of Experts on Phytosanitary Measures and Regional Plant Protection Organisation (RPPO) Technical Meetings at FAO United Nations in Rome every 2 years. He was also responsible for co-ordinating many other projects for SPC from 1993-1996.

The South Pacific Commission is now known as the Secretariat for the South Pacific Community.

He moved to Auckland, New Zealand in June 1996 and worked in various businesses.

In 2011, he started writing stories for children, novels, poetry, humor, religious and science books. He now writes full time and works as a volunteer for the Project Revival Charity Trust (Inc.). He also has other family projects.

Abbreviations used.

1. ABA- Abscisic acid
2. ACIAR - Australian Centre for International Agriculture Research.
3. ACP - Africa, Caribbean, Pacific.
4. AusAID - Australian AID.
5. BAP - 6-Benzylaminopurine
6. BATF - Biosecurity and Trade Support.
7. BATS - Biosecurity Trade Support.
8. BIF - Biosecurity Information Facility.
9. CABI - Commonwealth Agriculture Bureau International.
10. CePaCT - Centre for Pacific Crops and Trees.
11. CEPM - Committee of Experts on Phytosanitary Measures.
12. CGIARI - Consortium of International Agricultural Research.
13. EU - European Union.
14. FAO - Food and Agriculture Organisation of the United Nations.
15. FHIA - Honduras Foundation for Agricultural Research.
16. GCCA - Global Climate Change Alliance.

17. ICCAI - International Climate Change Adaptation Initiative.
18. IMPEXTEK - Import and Export Technology Centre.
19. INEA - International Network for Edible Aroids.
20. IPPC - International Plant Protection Commission.
21. IRA - Import Risk Assessment.
22. IRETA - Institute for Research, Extension and Training in Agriculture.
23. LN - Liquid Nitrogen.
24. LRD - Land Resources Division.
25. MAFF - Ministry of Agriculture, Fisheries and Forests.
26. MS - Murashige and Skoog
27. MSO - Murashige and Skoog(+sucrose)
28. NAA - Napthaleneacetic acid
29. NPPS - National Plant Protection Staff.
30. NPPO - National Plant Protection Organisations.
31. PARDI - Pacific Agribusiness Research for Development Initiative.
32. PICT - Pacific Island Countries and Territories.
33. PIPLD - Pacific Island Pest List

Database.

34. PNG - Papua New Guinea

35. PPPO - Pacific Plant Protection Organisation.

36. PRAP - Pacific Regional Agricultural Programme.

37. PSIS - Pacific Small Islands States.

38. RPPO - Regional Plant Protection Organisations.

39. SPC - Secretariat for the South Pacific Community.

40. SPC - South Pacific Commission.

41. TaroGen - Taro genetic Resources Conservation and Utilization.

42. TLB - Taro Leaf Blight.

43. UN - United Nations

44. USP - University of the South Pacific.

Literature and picture sources.

1. Flora.coa.gov.tw
2. Internet Pictures
3. International Atomic Energy Agency
4. Wikipedia
5. www.spc.int
6. www.fao.org

Members of the
South Pacific Commission.

The South Pacific Commission was established in 1947 by the Canberra Agreement after World War II as a 'security' organisation. Over the years, it increasingly got involved in other regional work and has become the premier technical organisation in the Pacific in areas of Agriculture, Demography, Health, Women's Development, Fisheries Research, Media, Community Education and Training and many others. There were 27 member countries in 1993. They include Australia, American Samoa, Cook Islands, Federated States of Micronesia, Fiji, France, French Polynesia, Guam, Kiribati, Mariana Islands, Marshall Islands, Nauru, New Zealand, New Caledonia, Niue, Palau, Papua New Guinea, Pitcairn Island, Samoa, Solomon Islands, Tonga, Tuvalu, United Kingdom, United States of America, Vanuatu, Wallis and Futuna.

University of the South Pacific...

The University of the South Pacific is the only regional university in the Pacific. It was established in 1968 to help train the young people of the Pacific Islands. It has 12 member countries. They are Cook Islands, Fiji, Kiribati, Marshall Islands, Nauru, Niue, Samoa, Solomon Islands, Tokelau, Tonga, Tuvalu and Vanuatu.

www.ingramcontent.com/pod-product-compliance
Lightning Source LLC
Chambersburg PA
CBHW060615210326
41520CB00010B/1343